Mathematical Snacks

A collection of interesting ideas
to fill those spare moments

Jon Millington

Tarquin

Turn spare moments into enthusiastic mathematical moments.

Develop a can-do attitude towards mathematical thinking!

Show that there is mathematics to be discovered wherever you look.

I should like to thank my wife, Pat, for all her help and support during the preparation of this book.

© 2019: Jon Millington
I.S.B.N: 978-1-899618-51-4
Design: Magdalen Bear
Printing: Printed in the UK

Tarquin
Suite 74, 17 Holywell Hill
St Albans AL1 1DT
United Kingdom
www.tarquingroup.com

'A snack is not intended to replace a full meal but is often just what is needed at the time!'

All 45 of these mathematical topics and ideas for the classroom are quick to set up and are genuinely stimulating and enriching.

They have been deliberately chosen to offer and allow interesting mathematical thinking in varied situations without spoiling future work on the direct curriculum. Additionally, they are bounded in a way to discourage the snack turning into a major meal and so spoiling the appetite!

Use them to fill spare moments in normal lessons and also for maths clubs and end of term special activities.

They can also provide a source of thought-provoking investigations for children who need some extra work to do while the rest of the class finishes an earlier assignment.

For each snack there are also suggestions for further work in the form of follow-on ideas. All the answers are given.

If you have enjoyed this book there may be other Tarquin books which would interest you. For example, 'Geometry Snacks', 'More Geometry Snacks' and 'The Number Detective' by Jon Millington, 'A Puzzle a Day' by Vivien Lucas and several other mathematical puzzle books. Tarquin books are available from bookshops, toy shops and gift shops or directly from the publishers website www.tarquingroup.com.

Alternatively, if you would like our latest information in paper or email form, please contact us by email: info@tarquingroup.com, through Twitter @tarquingroup or by post to Tarquin, Suite 74, 17 Holywell Hill, St Albans, AL1 1DT, United Kingdom.

Triangular Dominoes

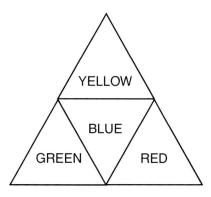

YELLOW

BLUE

GREEN   RED

This triangular domino has four different colours.

In how many different ways could such a shape be coloured using these four colours?

Follow-on Ideas

In how many different ways could it be coloured if two different colours are used?

How many if three different colours are used?

Mathematical Snack 1
Solutions & more ...

There are eight different ways of colouring such a domino.

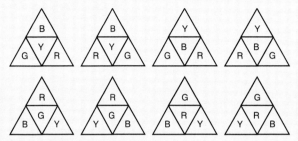

Follow-on Ideas

If two colours are used,
then there are six different ways.

```
   G          R          R          R          G          G
   R          G          R          G          R          G
G  G,      G  G,      G  G,      R  R,      R  R,      R  R.
```

If three colours are used,
then there are twelve different ways.

```
   R          G          B          G          B          R
   R          G          B          B          R          G
G  B,      R  B,      R  G,      R  R,      G  G,      B  B.
```

```
   R          G          B          B          R          G
   R          G          B          G          B          R
B  G,      B  R,      G  R,      R  R,      G  G,      B  B.
```

6

Magic Honeycomb

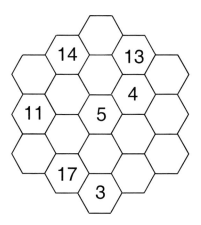

It is possible for the numbers from 1 to 19 to be placed in the hexagons in such a way that the total of any column or diagonal, whether it be 3, 4 or 5 hexagons long is always 38.

a. Fill in the missing numbers.

b. Is the honeycomb still magic if each number is doubled?

Follow-on Ideas

What happens if 2 is added or subtracted from every number? Does the honeycomb remain magic?

Try constructing another magic honeycomb with a different number of cells.

Mathematical Snack 2
Solutions & more ...

a.

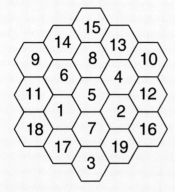

b. Yes.
Doubling each number does not affect the magic properties of this honeycomb even though different numbers of numbers are added together.

Yes. This is true of any multiplication factor.

Follow-on Ideas

No, the magic quality is lost. The total of a line depends on whether 3, 4 or 5 numbers are added together.

Number Patterns

Some multiplications give rise to fascinating patterns.
Here are some interesting ones to try.

a.
$$1 \times 9 + 2 =$$
$$12 \times 9 + 3 =$$
$$123 \times 9 + 4 =$$

and so on up to

$$12345678 \times 9 + 9 =$$

b.
$$1 \times 8 + 1 =$$
$$12 \times 8 + 2 =$$
$$123 \times 8 + 3 =$$

and so on up to

$$123456789 \times 8 + 9 =$$

Follow-on Ideas

Try multiplying the number 123456789
by the digits 2, 4, 5, 7, 8, in turn.
What is remarkable about all the answers?

What happens if it is multiplied by 3, 6 or 9?

Mathematical Snack 3
Solutions & more ...

The answers follow the patterns:

a.	b.
11	9
111	98
1111	987
11111	9876
111111	98765
1111111	987654
11111111	9876543
111111111	98765432
	987654321

Follow-on Ideas

x 2 = 246913578   x 4 = 493827156
x 5 = 617283945   x 7 = 864197523
x 8 = 987654312
All five include one of each digit except zero.

x 3 = 370370367   x 6 =740740734
x 9 =1111111101
These three introduce the digit 0 and
there are repetitions of various digits.

Snack 4

Magic Hexagon

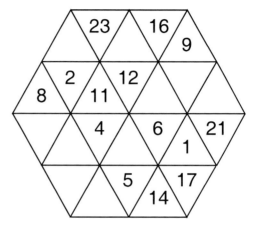

When this hexagon is filled in with the numbers from 1 to 24, once each, it is possible to obtain a magic total of 75 by adding any line of five or seven triangles.

Fill in the blank triangles.

Follow-on Ideas

If each number is doubled, the hexagon can be filled with the even numbers from 2 to 48 and the magic total will be 150.

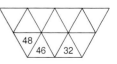

Reflect it so that the two largest numbers 48 and 46 are bottom left.

Mathematical Snack 4
Solutions & more ...

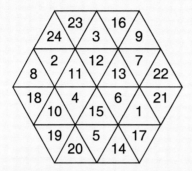

Follow-on Ideas

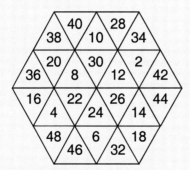

This is good practice in tracking a pattern and in doubling and reflecting at the same time.

All totals are 150.

Roman Cross Numbers

C M
D
X L V
I

All the answers in this cross number square are in the form of Roman numerals. Each is four letters long.

Across
1. 2 x 2 x 5 x 13
5. 2001 − 486
6. $4^2 + 3^2 + 1^2$
7. Two fifths of 155

Down
1. 20 x 47
2. 342 + 43 + 35
3. 528 ÷ 8
4. One third of one more than 50.

1	2	3	4
5			
6			
7			

Follow-on Ideas

1	2	3	4
5			
6			

Given that 5 across is 705 and 2 down is 1200, make up a Roman Cross Number for the rest.

Then make up suitable clues for them.

Mathematical Snack 5
Solutions & more ...

¹C	²C	³L	⁴X
⁵M	D	X	V
⁶X	X	V	I
⁷L	X	I	I

Across	Down
1. 260	1. 940
5. 1515	2 420
6. 26	3. 66
7. 62	4. 17

Follow-on Ideas

¹	²M	³	⁴
⁵D	C	C	V
⁶	C		

There are a few restrictions such that the top left square can only contain C or M, but in general anything goes!

Quadrilateral Discovery

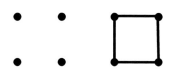 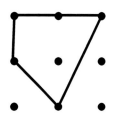

By joining the points of a 2 x 2 grid, it is only possible to draw just one quadrilateral, a square.

How many different quadrilaterals can be drawn on a 3 x 3 grid? Be careful not to include equivalent quadrilaterals that are the same shape and size as an earlier one but differently placed on the grid.

Follow-on Ideas

Try working with other grid sizes, such as 3 x 4 and 4 x 4.

Mathematical Snack 6
Solutions & more ...

There are sixteen possible quadrilaterals.

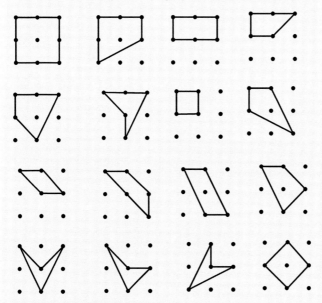

Follow-on Ideas

The number of possible shapes increases considerably with more points and it is best to regard this as open work for a few enthusiasts. Let them check the work of others!

Snack 7

Abacus Beads

This diagram shows a simple abacus with nine beads representing the number 324.

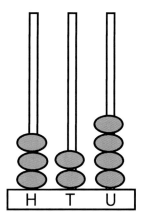

a. If there are twenty beads and all must be used, what is the largest number that can be represented?

b. What is the smallest number that can be represented?

c. If there are almost the same number of beads on each column, what numbers can be represented?

Follow-on Ideas

Write down, in size order, all the numbers that can possibly be represented by using all twenty beads.

Mathematical Snack 7
Solutions & more ...

a. The largest is 992.
b. The smallest is 299.
c. If the columns each have almost the same number, then there must be two sevens and a six.
   Hence 677, 767, 776.

Follow-on Ideas

Such problems encourage logical and systematic work.

The numbers are:
299
389 398
479 488 497
569 578 587 596
659 668 677 686 695
749 758 767 776 785 794
839 848 857 866 875 884 893
929 938 947 956 965 974 983 992

Snack 8

Repeated Names

If the name Maureen is written over and over again like this:

MAUREENMAUREENMA...

a. What is the 16th letter?

b. What is the 39th letter?

c. What is the hundred and fifth letter?

Follow-on Ideas

An interesting discussion could arise if the teacher suggests guessing what an unspecified letter in the sequence could be.

a. What is the probability of it being a U?

b. What is the probability of it being an E?

Once it is known that the unspecified letter is the 2722nd, what is the probability in each case?

Consider using names such as ANNA or GEORGETTE with several repeated letters. However, there will almost certainly be some interesting names to use within your actual class.

Mathematical Snack 8
Solutions & more ...

    a. A
    b. R
    c. N

This is really a kind of modular arithmetic. The letters repeat in groups of seven. Whole groups of 7 can be eliminated.

$$16 = (2 \times 7) + 2 \longrightarrow 2 \longrightarrow A$$
$$39 = (5 \times 7) + 4 \longrightarrow 4 \longrightarrow R$$

It is a good idea to check any formula like this near the beginning

$$2 = (0 \times 7) + 2 \longrightarrow 2 \longrightarrow A$$
$$7 = (1 \times 7) + 4 \longrightarrow 4 \longrightarrow R$$

Where the number is a multiple of 7

$$105 = (15 \times 7) + 0 \text{ which is confusing, so write it as}$$
$$105 = (14 \times 7) + 7 \longrightarrow 7 \longrightarrow N$$

Follow-on Ideas

a. The probability is $1/7$ if the number of places is unknown.
b. The probability is $2/7$ if the number of places is unknown.
However, $2722 \longrightarrow 6 \longrightarrow E$.
Hence the probability of U is zero and of E is 1.

Snack 9

Number Sorting

$$0\ 1\ 2\ 3\ 4\ 5\ 6\ 7\ 8\ 9\ 10$$
$$11\ 12\ 13\ 14\ 15\ 16\ 17$$

Group A	Group B	Group C
0 3 6 8	1 4 7	2 5 9 10
	11	12 13 14

The numbers shown on the blackboard above have been divided into three groups according to some rule or principle. The table shows the numbers 1 to 14 in their correct groups.

Where should 15, 16 and 17 be put?

Follow-on Ideas

An English and a French Mathematician were each asked to sort the numbers from 1 to 12 into five groups using the same rule. These are their results.

	A	B	C	D	E
English		1, 2, 6	4, 5, 9	3, 7, 8	11, 12
French	1	6	2, 5, 7, 8, 9, 11	3, 12	4

Where should they each put 10, 20, and 30?

Mathematical Snack 9
Solutions & more ...

This question depends on the handwriting or the typeface of the numbers.

In the case of this example, shown on the blackboard, the numbers in group A have only curves, those in group B only straight lines and those in group C are mixed.

Hence 15 and 16 go into Group C and 17 into Group B.

Follow-on Ideas

This time the sorting is done according to the number of letters in the word for the number in each of the languages
The numbers of letters in the groups are:
(A) 2   (B) 3   (C) 4   (D)5   (E) 6

TEN, goes into Group B and TWENTY and THIRTY both go into Group E.
DIX goes into Group B, VINGT into Group D and TRENTE into Group E.

French numbers from 1 to 12 are:
un, deux, trois, quatre, cinq, six,
sept, huit, neuf, dix, onze, douze.

Groups of Numbers

1 2 3 4 5 6 7 8 9 10 ...

1st group	1	2
2nd group	3	

If you put whole numbers, starting from 1, into two groups, how far can you get so that no two numbers and their total appear in the same group?

For example, if you decided to put 1 and 2 in the first group, then 3 would have to go in the second, and so on.

Follow-on Ideas

How far can you get if there are three groups of numbers instead of two?

Mathematical Snack 10
Solutions & more ...

You can get as far as 8 with two groups.

First Group: 1, 2, 4, 8.
Second Group: 3, 5, 6, 7.

Follow-on Ideas

You can get as far as 23 with three groups.

First Group: 1, 2, 4, 8, 11, 22.
Second Group: 3, 5, 6, 7, 19, 21, 23.
Third Group: 9, 10, 12, 13, 14, 15, 16, 17, 18, 20.

Counting the Days

M	Tu	W	Th	F	Sa	Su
1	2	3	4	5	6	7

a. Is it possible to have five Sundays in the same month?

b. Is it possible for February to have five Sundays?

c. Is it possible for December to have six Sundays?

d. If the 19th of a month is a Sunday, what date is the last Monday of the month?

e. Can there ever be more Wednesdays than Mondays in the same month?

f. In the past, many people were paid on Fridays. Although there are 52 weeks in a year, could there ever be 53 pay days.

Follow-on Ideas

If you are looking for a really exceptional month, then September 1752 is the one to choose.

M	Tu	W	Th	F	Sa	Su
		1	2	14	15	16	17
18	19	20	21	22	23	24
25	26	27	28	29	30	

It was the month when the Gregorian calendar was introduced in Britain and America. Eleven days were missed out, so Tuesday September 2nd was followed by Wednesday September 14th. It was a very short month!

25

Mathematical Snack 11
Solutions & more ...

a. Yes.
b. Yes - as long as it is a leap year.
c. No - to get 6 Sundays in requires a minimum of 36 days and there aren't any such months.
d. The 27th. There cannot be a Monday after that.
e. Yes - as long as the month begins on a Tuesday or a Wednesday.
f. 52 weeks is 364 days, so if January 1st is a Friday, there can be 53 Fridays. A leap year has 366 days, so whenever such a year begins on either a Thursday or a Friday there are 53 Fridays.

Square Faces

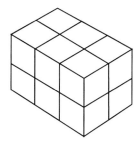

a. This diagram shows a cuboid made up of little cubes. How many are there?

b. How many cubes can you see part of?

c. How many cubes can you see no part of?

d. If you made all the cubes into a single column, standing on a table what is the largest number of square faces that you can see at one time?

Follow-on Ideas

This idea of making cuboids out of a number of small cubes could easily be extended to produce a whole series of investigations.

Given say 20, 30 or 40 cubes, make a cuboid with the largest or smallest number of visible (a) square faces (b) cubes.

Define a rule which will give an answer to each of these two questions when given the number of cubes in a particular cuboid.

Mathematical Snack 12
Solutions & more ...

a. 2 x 2 x 3 = 12 small cubes
b. 10 (so often people say 16)
c. Only 2
d. 12 + 12 + 1 = 25

This is a problem which can be posed either using a set of 12 cubes arranged in this way - or as a diagram on a board, an overhead slide or a photocopy.

Follow-on Ideas

As you can only ever see half a cuboid, the number of visible and invisible square faces will be the same.
For a cube measuring a x b x c, this number is ab + ac + bc and, the more the cuboid resembles a cube the more its value decreases.
For example with 40 cubes where abc = 40,
the number of visible square faces ranges from
81 (for 1 x 1 x 40) to 38 (for 2 x 4 x 5).

For a cube measuring a x b x c, the number of visible cubes is abc - ((a - 1)(b - 1)(c - 1)). As with the square faces this value decreases the more the cuboid resembles a cube.
For example with 40 cubes where abc = 40,
the number of visible cubes ranges from
40 (for 1 x 1 x 40) to 28 (for 2 x 4 x 5).

28

Snack 13

Pages in a Booklet

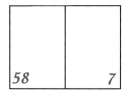 

A booklet is made up from single sheets of paper which are folded in half to create four pages of the booklet. The convention is for right-hand pages to have odd numbers and for left-hand pages to have even ones.

What are the two numbers on the other side of this sheet?

How many pages are there in the booklet and how many sheets are needed?

Follow-on Ideas

Another booklet, similarly made up, has 96 pages. What are the page numbers at the centre?
Which page comes on the same side of the same sheet as page 73?
What comment would you make about a sheet on which there are the four numbers 10, 11, 86, 87?

Mathematical Snack 13
Solutions & more ...

8 and 57
64 (one less than 7 + 58)
16 sheets

Follow-on Ideas

48 and 49
24 (97 - 73)
The booklet has been wrongly paginated.
9, 10, 87 and 88 would be the conventional
numbering.

Kaprekar's Constants

Choose any three figure number, except one where all the digits are the same. Rearrange the digits in descending order and subtract from this the number you get when the digits are placed in ascending order.

Repeat this process with your answer and keep doing so until it produces the same sum over and over again.

What is the greatest number of subtraction sums you might have to do before you end up with an answer that would repeat itself for ever?

Follow-on Ideas

a. What happens if you start with a two-digit number?

b. What happens if you start with a four-digit number?

Mathematical Snack 14
Solutions & more ...

Five sums.
For example starting with 909 you get:

990	981	972	963	954
-099	-189	-279	-369	-459
891	792	693	594	495

D.R.Kaprekar published his discovery of the existence of these constants in an article 'An Interesting Property of the Number 6174' in *Scripta Mathematica* 21 (Dec. 1955) p.304.

Follow-on Ideas

a. You end up with one of these five: 9, 81, 63, 27, 45. They form a cyclic group, where each leads to the next, and the last one leads back to 9.

b. You will eventually end up with 6174.

Sign Language

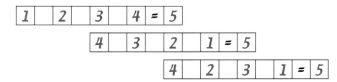

| 1 | | 2 | | 3 | | 4 | = | 5 |

| | 4 | | 3 | | 2 | | 1 | = | 5 |

| | | 4 | | 2 | | 3 | | 1 | = | 5 |

a. Use the signs + - x ÷ and doing the single calculations as you go, arrive at an answer of 5.

b. Now here is a similar puzzle which uses the nine digits 1 to 9 in a rather special way.

| 9 | | 8 | | 7 | | 6 | | 5 | | 4 | | 3 | = | 21 |

There are five different solutions.

c. Here is another puzzle arranged in the form of a square. Put signs in the right places so that it spirals in to an answer of 18.
There are nine different solutions.

9		4	
		5	= 3
2	1	8	
	6		7

Follow-on Ideas

Ask the class to construct their own version of the puzzle. You could say that the best possible have two or three solutions - not too many, nor too few. If they can use an interesting combination of digits, so much the better.

Mathematical Snack 15
Solutions & more ...

a. $1 + 2 \times 3 - 4 = 5$
$1 + 2 \div 3 + 4 = 5$

$4 + 3 - 2 \times 1 = 5$
$4 + 3 - 2 \div 1 = 5$
$4 \times 3 \div 2 - 1 = 5$

$4 - 2 + 3 \times 1 = 5$
$4 - 2 + 3 \div 1 = 5$
$4 - 2 \times 3 - 1 = 5$
$4 \times 2 - 3 \times 1 = 5$
$4 \times 2 - 3 \div 1 = 5$
$4 \div 2 + 3 \times 1 = 5$
$4 \div 2 + 3 \div 1 = 5$
$4 \div 2 \times 3 - 1 = 5$

b. $9 + 8 + 7 + 6 \div 5 \times 4 - 3 = 21$
$9 + 8 + 7 \div 6 \times 5 + 4 - 3 = 21$
$9 + 8 - 7 + 6 - 5 - 4 \times 3 = 21$
$9 + 8 - 7 - 6 \times 5 + 4 - 3 = 21$
$9 - 8 \times 7 - 6 + 5 \times 4 - 3 = 21$

c. $9 + 4 + 3 - 7 + 6 - 2 + 5 = 18$
$9 + 4 - 3 + 7 - 6 + 2 + 5 = 18$
$9 + 4 \times 3 - 7 - 6 \div 2 + 5 = 18$
$9 - 4 + 3 + 7 + 6 + 2 - 5 = 18$
$9 - 4 - 3 + 7 + 6 - 2 + 5 = 18$
$9 \times 4 + 3 - 7 - 6 \div 2 + 5 = 18$
$9 \times 4 - 3 + 7 + 6 \div 2 - 5 = 18$
$9 \times 4 \div 3 + 7 + 6 - 2 - 5 = 18$
$9 \times 4 \div 3 - 7 + 6 + 2 + 5 = 18$

Arrowgraphs of Digits

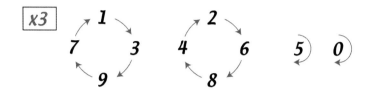

This diagram, called an arrowgraph, shows what happens to the units digits of numbers when they are multiplied by 3.
For example:
$2 \times 3 = \underline{6}$ $\underline{6} \times 3 = 1\underline{8}$ $\underline{8} \times 3 = 2\underline{4}$ $\underline{4} \times 3 = 1\underline{2}$ $\underline{2} \times 3 = 6$
which brings us back to where we started.
Any number ending in 5 will still end in 5 when multiplied by 3. Likewise for any number ending in 0.

Produce an arrowgraph for 7.

Follow-on Ideas

Construct arrowgraphs for 1, 2, 4, 5, 6, 8 and 9.

Mathematical Snack 16
Solutions & more ...

$\boxed{x7}$ 1 → 3 → 9 → 7   2 → 6 → 8 → 4   5 ↻   0 ↻

Follow-on Ideas

$\boxed{x1}$ 1 ↻   2 ↻   3 ↻   4 ↻   5 ↻   6 ↻   7 ↻   8 ↻   9 ↻   0 ↻

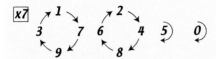

$\boxed{x2}$ 1 → 2 → 4 → 7 → ... → 6 → 8 → 3 → 9   5 → 0 ↻

$\boxed{x4}$ 1 → 4 → 6 → 9 → ... → 3 → 2 → 8 → 7   5 → 0 ↻

$\boxed{x5}$ 1 → 3 → 7 → 9 → 5 ↻   2 → 4 → 6 → 8 → 0 ↻

$\boxed{x6}$ 1 → 6 ↻   3 → 8 ↻   5 → 0 ↻   7 → 2 ↻   9 → 4 ↻

$\boxed{x8}$ 1 → 8 → 4 → 2 → ... → 6 → 7 → 9   5 → 0 ↻

$\boxed{x9}$ 1 → 9 ↻   2 → 8 ↻   3 → 7 ↻   4 → 6 ↻   5 ↻   0 ↻

Of course these are not the only diagrams that could be
used to illustrate this information. Designing clear diagrams
to show a particular property is always important.

Networks & Nodes

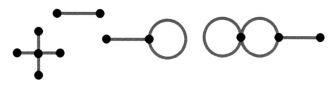

A network consists of lines called arcs which can be curved or straight. Their shape does not matter. What matters is the number of arcs and the number which meet at the nodes of the network. A free end of an arc is also considered as a node, and has a value of 1.

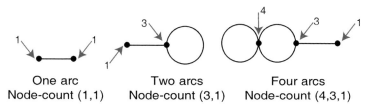

One arc	Two arcs	Four arcs
Node-count (1,1)	Node-count (3,1)	Node-count (4,3,1)

a. Find a second two-arc network with a different node-count.

b. Find the six possible networks with three arcs.

Follow-on Ideas

There are 14 four-arc networks including the one above and certain classes will certainly enjoy trying to find them all. The total node-count is 8 in each case.

It is worth pointing out that a given node-count may produce more than one distinct diagram.

Mathematical Snack 17
Solutions & more ...

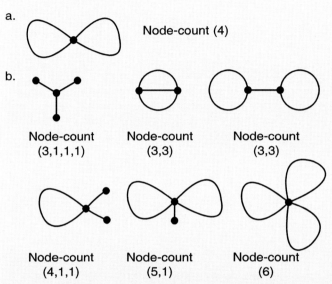

a.

Node-count (4)

b.

Node-count
(3,1,1,1)

Node-count
(3,3)

Node-count
(3,3)

Node-count
(4,1,1)

Node-count
(5,1)

Node-count
(6)

The main problem is recognising and identifying topological equivalents to these diagrams. Encourage simplicity.

Follow-on Ideas

The 14 four-arc networks have the following node-counts:
(3,3,1,1) (3,3,1,1) (4,1,1,1,1) (4,3,1) (4,3,1) (4,3,1)
(4,4) (4,4) (5,1,1,1) (5,3) (5,3) (6,1,1) (7,1) (8)

There are 39 five-arc networks to discover!
Remember that each node-count will be 10.

Clock Face Primes

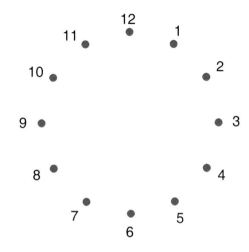

Draw a clock face and join each prime number to all the other primes. Remembering that 1 is not a prime number, how many different lengths of line have you drawn?

Follow-on Ideas

Clock faces such as this offer many possibilities. Another interesting one is to ask the same question about lengths when each even number is joined to every other.

In this case how many different regions have been created?

Mathematical Snack 18
Solutions & more ...

There are six different lengths.

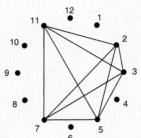

2 - 3

3 - 5 & 5 - 7

2 - 5 & 2 - 11

3 - 7 & 3 - 11 & 7 - 11

2 - 7

5 - 11

Follow-on Ideas

There are three different lengths.

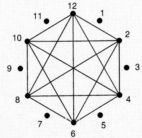

12 - 2

12 - 4

12 - 6

There are 24 regions,
four in each of the six
equilateral triangles.

Ten Trees to Plant

Ten trees can be planted in such a way that four of them lie in each of the five rows.

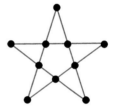

Can you find other ways of planting them so that there are four in each of the five rows? All the rows must be straight. (There are in fact five more ways to discover.)

Follow-on Ideas

What about twelve trees in six straight rows?
To limit the number of solutions, trees can only be at intersections and every intersection must have a tree.

Mathematical Snack 19
Solutions & more ...

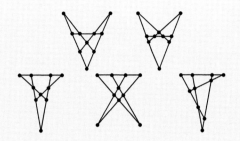

Follow-on Ideas

With the constraints mentioned, there are only six solutions.

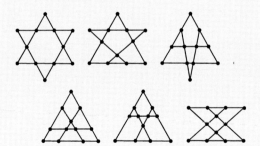

Snack 20

Triangular Numbers

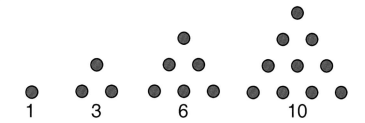

1   3   6   10

The diagram shows the first four triangular numbers. You can see that you get the second one by adding 2 to the first, and the third one by adding 3 to the second, and so on.

If you keep going you will meet two consecutive triangular numbers with the same units digit. What are they?

Rather further on, you will some across two more consecutive numbers with the same units digit. What are they?

Follow-on Ideas

An interesting discussion could arise once someone has spotted that the first set are the 9th and 10th and the second set the 19th and 20th. Confirm that the 29th and 30th will have the same property.

The idea arises naturally of how to calculate any given triangular number. Depending on the class, the formula $n(n + 1) \div 2$ or a more verbal way of describing how to work them out can be discussed. What are the 99th and 100th triangular numbers?

Mathematical Snack 20
Solutions & more ...

45 and 55
190 and 210

Follow-on Ideas

435 and 465
4950 and 5050

Starting with the 9th triangular number, then the 19th, 29th, and so on, you will be adding a multiple of 10 to obtain the next triangular number. The ending of such pairs alternates between 5 and 0.

Anagram Arrangements

A E R

How many different arrangements of the three letters above can be found?

How many of them are real words?

O P S T

How many different arrangements of the four letters above can be found?

How many of them are real words?

Follow-on Ideas

Another useful set of four letters which produce five real words is

A E S T

With five letters there are 120 arrangements and

A E L S T

is a useful one to try.

Mathematical Snack 21
Solutions & more ...

With A E R, there are six arrangements and the real words are:
ARE, EAR, ERA.

With O P S T there are twenty-four arrangements with six real words:
OPTS, POST, POTS, SPOT, STOP, TOPS.

Follow-on Ideas

With A E S T there are twenty-four arrangements with five words:
EAST, EATS, SATE, SEAT, TEAS.

With A E L S T there are 120 arrangements with six real words:
LEAST, SLATE, STALE, STEAL, TALES, TEALS.

Snack 22

Rectangles in a Rectangle

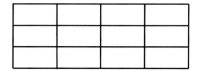

How many rectangles are there altogether in this diagram, not forgetting the surrounding rectangle?

Follow-on Ideas

What happens to the total number of rectangles if you add another row along the bottom, changing the 3 x 4 shape of the figure to 4 x 4?

What happens to the total number of rectangles if you add another row along the left side, changing the 3 x 4 shape of the figure to 3 x 5?

Mathematical Snack 22
Solutions & more ...

WIDTH

		1	2	3	4
	1	12	9	6	3
HEIGHT	2	8	6	4	2
	3	4	3	2	1

This table shows the number of rectangles of each size, so the total is:

$$24 + 18 + 12 + 6 = 60.$$

Follow-on Ideas

1	2	3	4
16	12	8	4

For the 4 x 4 shape, add this row on top of the table for the 4 x 3 shape. There are 40 additional rectangles, making 100 in all. If it is not clear why this should be, start again with a 4 x 4 diagram.

1	15
2	10
3	5

For a 3 x 5 shape, add this column on the left. This adds a further 30 rectangles, giving a total of 90.

Reflected Times

The ten digits used in a digital clock are shown above. Suppose that a twelve-hour digital clock, in a coach, can be seen reflected in a side window. For example a 2 appears as a 5 and vice-versa. At what times would the reflection show the same time as the clock itself?

The clock always shows four digits, two for the hours and two for the minutes.

Follow-on Ideas

At what times would the reflections appear the same if it were a twenty-four hour digital clock and always shows four digits?

Mathematical Snack 23
Solutions & more ...

The clock and its reflection would show the same times at these six times:

01:10   02:50   05:20

10:01   11:11   12:51

Follow-on Ideas

The above six plus the five below making eleven times in all.

00:00   15:21   20:05

21:15   22:55

Giving Change

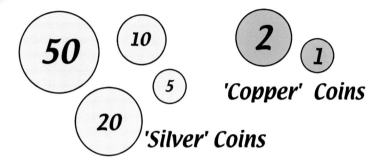

'Copper' Coins

'Silver' Coins

The coins above offer the simplest and most efficient way of reaching all possible sums of money and so it is used in most countries of the world.

In how many different ways could a 10 unit coin be exchanged for others?

Follow-on Ideas

In how many different ways can a purchase of value 50 units be paid for using only 'silver' coins?

Mathematical Snack 24
Solutions & more ...

There are 10 different ways.

```
5  5
5  2  2  1
5  2  1  1  1
5  1  1  1  1  1
2  2  2  2  2
2  2  2  2  1  1
2  2  2  1  1  1  1
2  2  1  1  1  1  1  1
2  1  1  1  1  1  1  1  1
1  1  1  1  1  1  1  1  1  1
```

Encourage systematic and orderly listing of possibilities.

Follow-on Ideas

There are 13 different ways.

```
50
20 20 10
20 20  5   5
20 10 10 10
20 10 10  5   5
20 10  5   5   5   5
20  5   5   5   5   5   5
10 10 10 10 10
10 10 10 10  5   5
10 10 10  5   5   5   5
10 10  5   5   5   5   5   5
10  5   5   5   5   5   5   5   5
 5   5   5   5   5   5   5   5   5   5
```

Points on a Circle

| 2 points | 3 points | 4 points |
| 2 regions | 4 regions | 8 regions |

By marking equally-spaced points on a circle and then joining each to all others, you create regions. In the three examples above, the process creates 2, 4, 8 regions.

a. How many regions are there with 5 points?

b. How many regions are there with 6 points?

Follow-on Ideas

Investigate what happens when the points are not evenly spaced.

Mathematical Snack 25
Solutions & more ...

a. 5 points give 16 regions.
b. 6 points give 30 regions.
   (Not 32 as so many people guess!)

Follow-on Ideas

If the points are not evenly spaced, there is no change up to 5 points. For 6 points, there might or might not be an extra triangular area in the middle, depending on the location of the points.

Striking Clocks

If a grandfather clock strikes the number of hours at each hour, how many strikes would there be in a day?

Follow-on Ideas

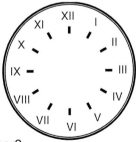

Roman striking clocks have two bells. Each I is struck on the high bell and each V on the low one. X counts as two V's and on this kind of clock 4 always appears as IV and not IIII.

How many strikes a day would you hear?

Mathematical Snack 26
Solutions & more ...

There are twice $1 + 2 + 3 + ... + 12 = 156$ strikes.

Follow-on Ideas

There are 4 X's which will sound as 8 V's,
there are 5 real V's and 17 I's,
so a total of $8 + 5 + 17 = 30$.
Double this for a complete day to give 60 strikes.

This method of striking has not been made up for this
question! Joseph Knibb made several clocks, both
longcase and bracket, using this system. This was around
1680 and was an attempt to conserve spring power and so
reduce the amount of winding required.

Number Chains

$$25 \rightarrow 24 \rightarrow 12 \rightarrow 6 \rightarrow ?$$
$$38 \rightarrow 19 \rightarrow 18 \rightarrow 9 \rightarrow ?$$

Choose any whole number. Halve it if it is even or subtract 1 if it is odd. Create a number chain by repeating this process over and over again.

The two number chains above show how the chains for 25 and for 38 start. Continue them until something surprising happens. How many steps did it take?

Try a succession of other starting numbers. What happens now?

Follow-on Ideas

Form number chains in a way similar to those above by halving even numbers but this time when the number is odd, treble it and add 1. Here is what happens to 22.

$$22 \rightarrow 11 \rightarrow 34 \rightarrow 17 \rightarrow 52 \rightarrow 26 \rightarrow ?$$

What happens eventually? Now try this rule on a succession of other numbers.

What happens if the rule to halve even numbers remains but the rule for odd numbers becomes 'double it and add 1'. What if the rule for odd numbers is 'multiply by 5 and add 1' ?

Mathematical Snack 27
Solutions & more ...

You always end up with zero.

25 ➜ 24 ➜ 12 ➜ 6 ➜ 3 ➜ 2 ➜ 1 ➜ 0

38 ➜ 19 ➜ 18 ➜ 9 ➜ 8 ➜ 4 ➜ 2 ➜ 1 ➜ 0

The first chain takes 7 steps to get to zero and the second takes 8 steps.

Follow-on Ideas

22 ➜ 11 ➜ 34 ➜ 17 ➜ 52 ➜ 26 ➜ 13 ➜ 40
➜ 20 ➜ 10 ➜ 5 ➜ 16 ➜ 8 ➜ 4 ➜ 2 ➜ 1

This new rule requires more steps, in this case 15, but you always end up with 1.

A starting point of 56 generates a number chain with 19 steps.

Doubling and adding one produces a sequence that dashes off to infinity. Multiplying by 5 and adding 1 produces unacceptably long chains. Trebling and subtracting 1 can produce repeating loops.

Just Four Make Twenty

1	2	3
4	5	6
7	8	9

1	2	3
4	5	6
7	8	9

The diagram above shows the digits from 1 to 9 written in a 3 x 3 square.

If the four corners are shaded and the numbers added together, their total is 20. However, this is not the only way to produce a total of 20 from the numbers in four of the squares.

Each on a separate diagram, shade or colour all the different ways that four squares can produce a total of 20. How many can you find?

Follow-on Ideas

Try this with a different set of nine consecutive numbers in the square, say from 10 to 18.

The total should then be 56(20 + 4 x 9).

How is the number in the centre square related to this total?

Mathematical Snack 28
Solutions & more ...

There are 11 more patterns, making 12 in all.

1	2	3
4	5	6
7	8	9

1	2	3
4	5	6
7	8	9

1	2	3
4	5	6
7	8	9

1	2	3
4	5	6
7	8	9

1	2	3
4	5	6
7	8	9

1	2	3
4	5	6
7	8	9

1	2	3
4	5	6
7	8	9

1	2	3
4	5	6
7	8	9

1	2	3
4	5	6
7	8	9

1	2	3
4	5	6
7	8	9

1	2	3
4	5	6
7	8	9

1	2	3
4	5	6
7	8	9

Follow-on Ideas

The patterns are the same with any set of nine consecutive numbers and the central number will always be a quarter of the total of those in the four corners.

Factor Trees

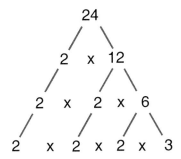

To make a factor tree, write down the number to be factorised and beneath it put the lowest prime that divides into it and the number of times it goes.

Continue until you have a row containing only primes, as shown here for 24.

Now produce factor trees for 60, 72 and 100.

Follow-on Ideas

Factor trees are another way of approaching the square roots of numbers that are perfect squares. The final row will contain even numbers of all the prime factors and the square root can be identified.

Make factor trees of 196 and 324 and find their square roots. Then 216 and find its cube root.

Mathematical Snack 29
Solutions & more ...

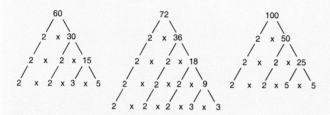

Follow-on Ideas

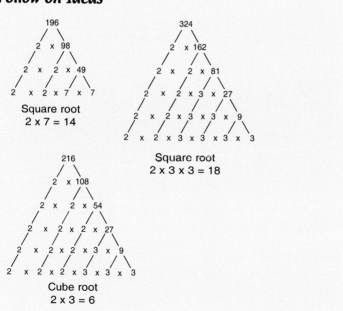

Clock Angles

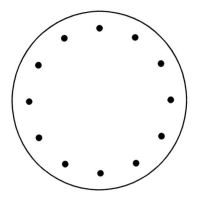

These twelve dots represent the hours on a clock face. What is the acute angle in degrees between the line joining the dots for 2 and 9 and the line joining the dots for 3 and 11?

Follow-on Ideas

What is the acute angle between the line joining 10 and 4 and the line joining 9 and 1?

Mathematical Snack 30
Solutions & more ...

Draw a line from 4 to 10 which is parallel to the one from 3 to 11.

Then draw the line from 1 to 10, which is parallel to the one from 2 to 9.

Join 1 to 4.

The dashed triangle is right-angled and isosceles and so the required angle is 45°.

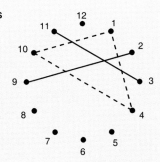

Follow-on Ideas

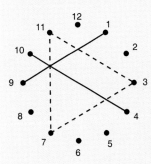

Draw a line from 3 to 7 which is parallel to the one from 1 to 9.

Then draw the line from 3 to 11, which is parallel to the one from 4 to 10.

Join 7 to 11.

The dashed triangle is equilateral and so the required angle is 60°.

Shading Squares & Circles

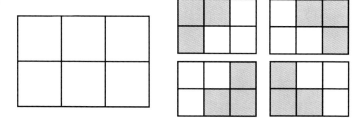

A 2 x 3 rectangle is divided into six squares and the object is to shade half of it, shading only whole squares. How many different such shadings are there?

Note that the four smaller rectangles show only one way because they are all reflections or rotations of each other.

Follow-on Ideas

a. This circle is divided into eight sectors. In how many ways can it be shaded so that exactly four of the sectors are shaded?

b. In how many different ways can three out of the eight sectors be shaded?

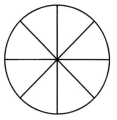

Mathematical Snack 31
Solutions & more ...

These are six distinct arrangements, ignoring reflections and rotations.

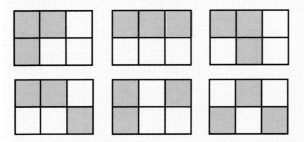

Follow-on Ideas

There are eight distinct arrangements, ignoring reflections and rotations, where four sectors are shaded.

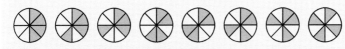

There are five distinct arrangements, ignoring reflections and rotations, where three sectors are shaded.

Creating Sequences

A number sequence begins

2, 6, 15 ...

The first term is $1 + 1^2 = 2$ and the sequence then continues by adding successive squares to the previous term.

So $2 + 2^2 = 6$, $6 + 3^2 = 15$, $15 + 4^2 = 31$, and so on.

Find the next three terms and the tenth term.

Follow-on Ideas

How many other ways can you find of creating the sequence 2, 6, 15 and then continuing it?

Mathematical Snack 32
Solutions & more ...

$31 + 5^2 = 56$, $56 + 6^2 = 92$, $92 + 7^2 = 141$.

The tenth term is $141 + 8^2 + 9^2 + 10^2 = 386$.

The nth term is the sum of the
first n square numbers plus 1, which is
$n(n + 1)(2n + 1) \div 6 + 1$.
So 10th term is $10 \times 11 \times 21 \div 6 + 1 = 386$.

Follow-on Ideas

This is a very good investigation.
Here are five possible continuations:

a. **2** (+ 4)   **6** (+ 4 + 5) **15** (+ 4 + 5 + 6)   **30** (+ 4 + 5 + 6 + 7) **52**

b. **2** (x 2 + 2) **6** (x 2 + 3) **15** (x 2 + 4)   **34** (x 2 + 5)   **73**

c. **2** (+ 4)   **6** (+ 4 + 5) **15** (+ 4 + 5 + 5)   **29** (+ 4 + 5 + 5 + 5) **48**

d. **2** (x 3)   **6** (x 2.5) **15** (x 2)   **30** (x 1.5)   **45**

e. **2**   **6**   **15**   **40**   **104**
   (1 x 2)   (2 x 3)   (3 x 5)   (5 x 8)   (8 x 13)

(terms in the Fibonacci sequence)

No doubt there are many other ingenious and defendable
answers.

Matchstick Triangles

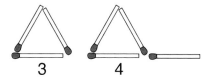

With three or four matchsticks it is only possible to make one triangle.

With five, however, it is possible to make two different triangles.

Sketch all the different triangles that can be made with seven matches. (You do not have to use all the matches each time.)

Follow-on Ideas

How many matchsticks are needed to make the same number of different triangles as there are matches? Not all of them need to be used each time.

Mathematical Snack 33
Solutions & more ...

Five different triangles can be made with seven matches.

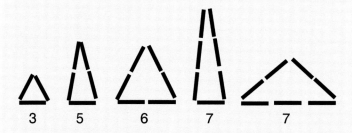

3   5   6   7   7

Follow-on Ideas

Nine matches can be used to produce nine triangles.

Here are the other four:

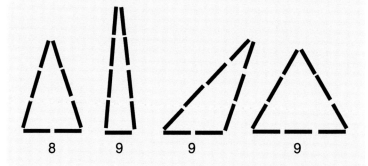

8   9   9   9

Cutting Gold Leaf

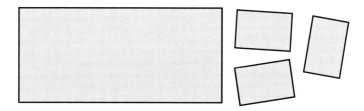

Gold leaf is very expensive and must not be wasted.

So the problem here is to decide how a rectangular piece measuring 5 cm x 11 cm can be cut to provide the maximum number of rectangles which are 2 cm x 3 cm?

How many such rectangles can be cut and what shape and area remains?

Follow-on Ideas

If the rectangles need to be 5 cm x 3 cm and the sheet is 14 cm x 11 cm, how many pieces can be cut and what shape and area remains?

Ask the class to make up similar questions to offer to a neighbour.

Mathematical Snack 34
Solutions & more ...

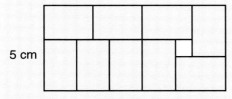

5 cm

11 cm

Nine pieces can be cut using the cutting plan above.
A 1 cm x 1 cm square remains. Area = 1 cm²

Follow-on Ideas

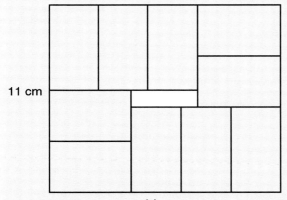

11 cm

14 cm

Ten pieces can be cut using the cutting plan above.
A 1 cm x 4 cm rectangle remains. Area = 4 cm²

Zero Some Digits

Snack 35

Replace as many of the digits 5, 6, 7, 8 with a zero as are necessary to make this addition correct.

```
  5 5 5
  6 6 6
  7 7 7
  8 8 8
-------
1 1 1 1
```

Follow-on Ideas

```
  5 5 5
  7 7 7
  8 8 8
  9 9 9
-------
1 1 1 1
```

Do the same with this problem. As with the one above, the solution is unique.

You can make up more of this kind of sum with different sets of three digits but in general there will be multiple solutions.

Mathematical Snack 35
Solutions & more ...

The only solution:

```
  5 0 5
  6 0 6
  0 0 0
  0 0 0
-------
1 1 1 1
```

It is possible to write a program in QBasic and no doubt in other languages to prove that this is the only solution.

Follow-on Ideas

The only solution:

```
  0 5 5
  0 7 7
  8 8 0
  0 9 9
-------
1 1 1 1
```

Starting the Millennium

Saturday January 1 2000

The first day of the year 2000 was a Saturday.

The puzzle is to work out on which day of the week the next millennium will start.

Under the Gregorian calender, leap years occur in those years when the final two digits are divisible by four. However the century years ending in 00 are not leap years unless the first two digits are also divisible by four. That is why the year 2000 was a leap year and why 3000 will not be.

Follow-on Ideas

Assuming the Gregorian Calendar is still in use, on which day of the week will January 1 fall in the years 4000 and 5000?

Is there a general rule you could apply?

Mathematical Snack 36
Solutions & more ...

By January 1 3000, exactly 1000 years will have passed, so if there were no leap years there would be 365 000 days.

If leap years came every four years without exception there would be 250 more days, starting in 2000 and ending in 2996.

However, the century years 2000, 2100, 2200 etc. are not leap years unless the first two digits are divisible by four, so we need to subtract 10 and then add 3 for the years 2000, 2400 and 2800, giving 365 243 days = 52 177 weeks and 4 days.
The next millennium therefore begins on a Wednesday.

Follow-on Ideas

For the millennium from 3000 to 4000, the number of days is 365 240 plus one extra day for each of 3200 and 3600, giving 365 242 days = 52 177 weeks and 3 days. The next millennium therefore begins on a Saturday.

For the millennium from 4000 to 5000, the number of days is 365 240 plus one extra day for each of 4000, 4400 and 4800, giving 365 243 days = 52 177 weeks and 4 days.
The next millennium therefore begins on a Wednesday.

The exceptional days will occur in 5200, 5600, and then 6000, 6400 and 6800. Millennium days will alternate between Saturdays and Wednesdays for as long as the present system remains in use!

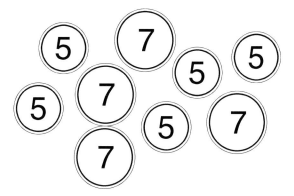

Token Payments

Given a large number of tokens of 5 and 7 units, it is easy to see how to pay for something exactly which costs 5, 7, 10, 12, 14 etc. units. It is also easy to see that it is not possible to pay exactly for something costing 1, 2, 3, 4, 6 or 8 units.

The puzzle is to find the largest amount that cannot be paid exactly using these tokens.

Follow-on Ideas

Trying other pairs of tokens with values such as 3 and 8 units can help to suggest a formula to calculate the largest total that you cannot pay exactly. What happens if the pair of values share a common factor?

Mathematical Snack 37
Solutions & more ...

Trial and error will produce a largest value of 23 units. After that there are five consecutive amounts which can be paid exactly.

24 = 5 + 5 + 7 + 7
25 = 5 + 5 + 5 + 5 + 5
26 = 5 + 7 + 7 + 7
27 = 5 + 5 + 5 + 5 + 7
28 = 7 + 7 + 7 + 7

All larger amounts can now be reached by adding one or more 5-unit tokens. Once seven consecutive amounts are found, larger amounts can also be reached by adding further 7-unit tokens.

Follow-on Ideas

For tokens of value 3 and 8 units, the largest total that you cannot make is 13 units. This is 3 x 8 - 3 - 8.

The value of 23 in the solutions above is 5 x 7 - 5 - 7.

The general formula is ab - a - b.

Where the two values have a common factor, all totals that can be reached must be a multiple of that factor. This means that there is no greatest value that cannot be reached. Try 3 and 6 and then 8 and 12.

Climbing to the Summit

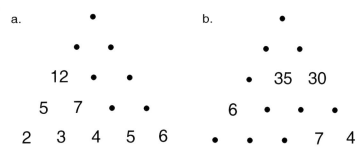

a.

```
        •
      •   •
  12  •   •
  5  7  •   •
  2  3  4  5  6
```

b.

```
          •
        •   •
    •  35  30
  6  •   •   •
  •   •   •   7   4
```

By adding the numbers in any row together in pairs you can find the numbers in the row above.
In example (a) 5, 7 and 12 have been filled in for you.
Complete the triangle to find the number at the top.

In example (b) apply the same rule and find
the number at the summit.

Follow-on Ideas

This one offers rather more of
a challenge and it is best to
introduce algebra as the way
to approach it.

Further examples can be
constructed or the class can be
asked to make them up for
others to solve.

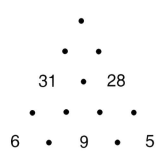

```
        •
      •   •
  31    •    28
 •    •   •    •
6   •   9   •   5
```

Mathematical Snack 38
Solutions & more ...

a.
```
            64
         28   36
      12   16   20
     5   7   9   11
   2   3   4   5   6
```

b.
```
            122
         57   65
      22   35   30
     6   16   19   11
   2   4   12   7   4
```

Follow-on Ideas

Introducing algebra by substituting letters for one or more of the missing numbers is the best way to track how the pyramid develops.
For this possible approach:

x - 6 + 9 = 31 - x.   So x = 14
y - 5 + 9 = 28 - y.   So y = 12

```
              •
          •      •
      31     •      28
    x   31-x   28-y   y
  6   x-6   9   y-5   5
```

Complete solution
```
            125
         64   61
      31   33   28
     14   17   16   12
   6   8   9   7   5
```

80

Equal Areas

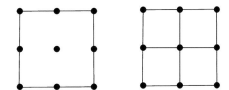

The diagram shows a lattice of nine equally-spaced dots drawn over a square. On the right one way of joining the dots with straight lines to make four equal areas has been shown. Not counting reflections or rotations, find all the other ways the square could be divided into four equal areas by joining the dots with straight lines.

Follow-on Ideas

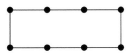

a. Another interesting arrangement to investigate is a 2 by 4 rectangle. With straight lines joining the dots, divide the rectangle into three equal areas.

b. Then try dividing the same rectangle into six equal areas.

Mathematical Snack 39
Solutions & more ...

There are ten solutions including the given one.

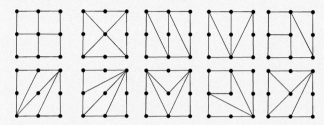

Follow-on Ideas

a. There are three solutions.

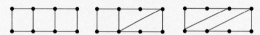

b. There are seven solutions.

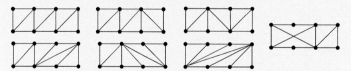

Such problems encourage logical and systematic work.

Adding in Pairs

2, 5, 7, 12, 19, . . .
8, 3, 11, 14, . . .

Here are two examples of the same process for generating a series of numbers. Start with any two numbers and write them side by side. Add them together and write the total on the right. Now add more terms to the series, with each new term being the sum of the previous two numbers.

In each case, how many times greater than the seventh number is the total of all ten?

Try the same process for any other pair of numbers.

Follow-on Ideas

Calling the first two numbers x and y, prove that the total of the first ten numbers will always be eleven times the seventh.

Mathematical Snack 40
Solutions & more ...

This is the same process as generates the Fibonacci series (which starts with 0 & 1). Sequences generated in this way but starting with other pairs of numbers are sometimes known as 'Lucas Numbers' after the French mathematician Edouard Lucas who studied them. It was he who first gave the name of 'Fibonacci numbers' to the sequence beginning 1, 1, 2, 3, 5, ...

In the first series, the seventh number is 50 and the total of all ten is 550, eleven times greater.

In the second series, the seventh number is 64 and the total of all ten is 704, eleven times greater.

Follow-on Ideas

The ten numbers will be:
x, y, $x + y$, $x + 2y$, $2x + 3y$, $3x + 5y$, $5x + 8y$,
$8x + 13y$, $13x + 21y$ and $21x + 34y$.
The total is $55x + 88y$ which is eleven times the seventh term $5x + 8y$.

Equilateral Triangles

a. Make as many different outlines as you can by joining four equilateral triangles edge to edge.
Don't count reflections or rotations.

b. Which outlines are also the net of a regular tetrahedron?

Follow-on Ideas

This diagram is one possible net for a square-based pyramid with equilateral triangles on four faces.

Not counting reflections or rotations, find all the other possible nets.

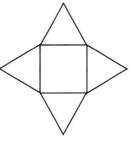

Mathematical Snack 41
Solutions & more ...

a. There are three outlines.

b. Just the first two nets.

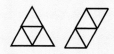

Follow-on Ideas

There are eight nets including the given one.

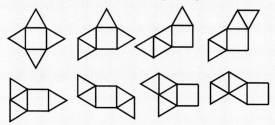

Maximum Products

$$a + b + c = 10$$
$$a \times b \times c = ?$$

What whole numbers would you choose so that they add up to 10 and can be multiplied together to make the largest possible product? For example, if the numbers had to add up to 4, they could be 1, 1, 1, 1 or 1, 1, 2 or 1, 3 or 2, 2. Clearly the last of these gives the largest product, 4.

Follow-on Ideas

Now try the same idea for numbers which add to 20. What is the largest possible product?

Mathematical Snack 42
Solutions & more ...

The numbers to choose are 2, 2, 3, 3,
giving a product of 36.

Follow-on Ideas

Contrary to what might be expected, the answer is
not 1296 which you get from multiplying two lots of
2, 2, 3, 3 together.

In fact it is 1458, which is 2 x 3 x 3 x 3 x 3 x 3 x 3.

Digit Multiplication Sums

(2, 3, 5, 7)   **5 x 5 = 25**
              5 x 7 = 35

Only the four digits 2, 3, 5 and 7 are allowed to be used in these multiplication sums, whether in the sum or in the answer. It is a fact that there are just three such sums where the answer is under 100.

Two are shown above. What is the third?

(2, 3, 5, 7)   **3 x 75 = 225**

Here a one-digit number is multiplied by a two-digit number to give a three-digit solution, once again using only the prime digits 2, 3, 5, 7. Find the other three multiplication sums that follow this pattern.

Follow-on Ideas

Keeping to the same restriction of using only the prime digits 2, 3, 5 and 7, investigate what happens when a single-digit number is multiplied by a three-digit number.

Mathematical Snack 43
Solutions & more ...

3 x 25 = 75

5 x 55 = 275
5 x 75 = 375
7 x 75 = 525

The digits 2,3,5,7 are the only single-digit primes.

Follow-on Ideas

There are four possible solutions:

7 x 325 = 2275
3 x 775 = 2325
5 x 555 = 2775
5 x 755 = 3775

There are no such sums with a three-digit product.

This idea can be extended by again using just 2, 3, 5, 7 to find other sums where a single-digit number is multiplied by a number with more than three digits.

Changing Places

a.
$$
\begin{array}{r}
76 \\
+\ 6 \\
\hline
70
\end{array}
$$

b.
$$
\begin{array}{r}
68 \\
\times\ 3 \\
\hline
17
\end{array}
$$

c.
$$
8\overline{)24}\ ^{71}
$$

In each of the sums above, all the digits are correct but they are in the wrong places. The puzzle is to move the digits so that each of the sums is correct.

Follow-on Ideas

In each of the multiplication sums below all the digits are correct but only the sevens are in the right places. Move the other digits so that the sums become correct.

d.
$$
\begin{array}{r}
693 \\
\times\quad 7 \\
\hline
681
\end{array}
$$

e.
$$
\begin{array}{r}
267 \\
\times\quad 2 \\
\hline
381
\end{array}
$$

f.
$$
\begin{array}{r}
495 \\
\times\quad 5 \\
\hline
8327
\end{array}
$$

Mathematical Snack 44
Solutions & more ...

a.
$$\begin{array}{r} 60 \\ + 7 \\ \hline 67 \end{array}$$

b.
$$\begin{array}{r} 13 \\ \times 6 \\ \hline 78 \end{array}$$

c.
$$4)\overline{72} = 18$$

Follow-on Ideas

d.
$$\begin{array}{r} 138 \\ \times 7 \\ \hline 966 \end{array}$$

e.
$$\begin{array}{r} 137 \\ \times 6 \\ \hline 822 \end{array} \left(\begin{array}{r} 227 \\ \times 3 \\ \hline 681 \end{array} \right.$$
This one must be eliminated because the 2 and 8 have not moved)

f.
$$\begin{array}{r} 583 \\ \times 9 \\ \hline 5247 \end{array}$$

Timers in Tandem

If you have two sand-filled timers, one for 11 minutes and one for 7 minutes, how would you use them to time 15 minutes without wasting any time at the start?

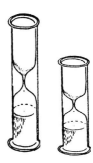

Follow-on Ideas

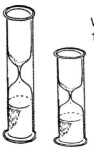

What would be the next length of time after 15 minutes that you could use the timers for?

Mathematical Snack 45
Solutions & more ...

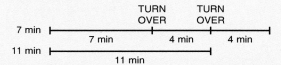

Start both timers. When the 7 runs out, turn it over. When the 11 runs out, turn the 7 over again. When the 7 runs out, the 15 minutes will be up.

Follow-on Ideas

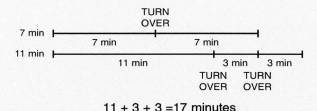

11 + 3 + 3 =17 minutes

The answer is not 18 minutes which you would get by starting the second timer when the first has run out, but 17 minutes.

*'A snack is not intended to replace
a full meal but is often just what
is needed at the time!'*

The Mathematical Snacks